I0390984

For those walking

shoulder-to-shoulder

next to someone

with schizophrenia

TOGETHER, THROUGH FLAMES

JASON VAN DONGEN

WITH

JACOBA VAN DONGEN

Together, Through Flames

> *True! — nervous — very, very dreadfully nervous I had been and am! But why will you say that I am mad? The disease had sharpened my senses — not destroyed — not dulled them. Above all was the sense of hearing acute. I heard all things in the heaven and in the earth. I heard many things in hell. How, then, am I mad? Harken! and observe how healthily — how calmly — I can tell you the whole story.*
>
> Edgar Allen Poe

In many ways, my childhood was a happy one. I have three younger siblings – a brother and two sisters – and we were brought up Christians. Although my creed would bring me both great comfort and great misery later in life, I still believe it is a fine way to live. The principles of love, tolerance and forgiveness are firm foundations on which to raise a family. Faith in God is no insurance policy against unhappiness, however. My mother – Cobi – has had mental health issues for as long as I can remember. Mum, would you like to briefly tell us about your struggles with your health?

My first experience of mental health issues occurred as a result of some ulcers I had in my mouth. I used to get them regularly, so when I was eight or nine, my mother took me to see a doctor. His opinion at the time was that it was caused by stress. It wasn't too long after, and I was on my first course of anti-depressants. I remember they were tiny red pills. I had a habit of soothing myself to sleep by imagining I was still a baby, but that I had different parents. I was afraid – very afraid – of my father. He used to punish us regularly, and sometimes I think he went overboard.

I didn't learn much at school. I had attention problems, and could never stay focused. I occasionally would burst into tears for no reason. This continued when I started work. When I was eighteen I was diagnosed with depression.

I had bad post-natal depression after all my children. I would also have to ask Dad to always check on you children to make sure you were all still there, and still breathing.

When I was about forty, I had several extended stays in hospital. Over the course of a year, I spent a total of seven months in hospital. While there, I experienced mild psychosis. I was afraid of monsters, and would bury myself under my sheets, but then I would hear them. I was very insecure at that time.

After I came home from hospital, I couldn't do much but sleep. I was also suicidal. One day, when nobody was home, I took thirty anti-psychotic pills and left a message with the pastor of our church to say good-bye to my husband and children. The pastor was there ten minutes later, but by then I was really confused, and I wondered why he was there. He took me to hospital. He saved my life, I think.

Since then, I have had a hospital admission virtually every year. I was in hospital when my first grandchild was born. It is only recently that I have avoided an annual hospital visit.

My own struggles with mental health started at a young age. I was perhaps eleven or so when school became a major problem for me. Over the course of one particular two-week term break, I was increasingly plagued by troubling dreams about school, and my anxiety regarding school rose. Come the start of the new term, I was a bundle of anxious nerves. I cried and wailed until my mother kept me home. I recall being a little surprised that my mother had been so easy to manipulate. The tears successfully kept me out of school for several weeks. Mum, what is your recollection of those days?

From the first, I knew it was because you couldn't cope. I'd try to encourage you to go to school, but I was in a bad place myself for some of the time.

As I recall, a doctor diagnosed me with depression shortly after. My mother sought out alternative therapies and remedies that might improve my health. Amongst other treatments, I had several sessions with a hypnotherapist. It failed to improve my mental health. The tears before school turned into tears in class. Over the course of a year, I went from being a cooperative student with good grades to being an underachiever with a poor attitude to schoolwork. Mum, did any of my teachers at that point in time express concerns about my mental health?

You had one teacher who advised us to toughen up. Then, in high school, you had a wonderful teacher who suggested we take you to see a doctor. You were diagnosed with depression, but the doctor didn't want to medicate you. Instead, we went to a hypnotherapist. I don't know if you remember this, but back then, you had already spoken about killing yourself. In fact, one of the reason we would let you pull out of school later on is we were genuinely

*afraid that if we pushed you too
hard, you would commit suicide.*

Time has a way of appearing shorter the further you stand from it. In hindsight, it appears my high school years flew by. Issues with my mental health – in particular, the crying during class – were amongst the reasons I pulled out of school before completing my eleventh year. I practically fell into fulltime work when a friend of my parents asked me to help set up a new store. Two managers and three years later, I quit my job and started collecting welfare. I met a beautiful young hell-raiser called Delta – my first ever girlfriend – and shuffled between rental homes until we landed in a small three bedroom house in a neighbourhood known for crime and violence, which had the effect of driving down the rental prices in the area. By this time, I had developed an unhealthy dependence on marijuana. I smoked copious amounts daily. Mum, do you remember if my behaviour changed after I developed my pot habit?

I remember being baffled how you and Delta could just sit there all day, on your furniture under your back patio, smoking pot, and not doing any housework or anything. During that time you also had some really eccentric ideas. You built castles in the sky, Jay, these great grand plans that were never ever going to actually happen.

Paranoia became a feature of my pot habit. For comfort, I would pick up the

Holy Bible. One day, blinded by tears, I picked up my Bible and it fell open to a Psalm in which this line is reiterated: 'For the mercy of the Lord endures forever.' All I could read between my tears was that line. It was a moment of momentous significance for me; I thought God had sent me a personal message. It changed how I perceived the Bible. Suddenly, I was a prophet, and the Bible was full of personal messages directed at me. Shortly after, I picked up a new fulltime job. Three months later, this came to an abrupt end when I delivered the manager what I thought was a message from Heaven – his store had been damned, and his corrupt business practises were coming to an end. God had spoken. Or so I thought. Mum, were you aware at the time that I was unwell? Is this something that was discussed?

I do not know at the time whether you heard voices, or were psychotic, but I remember you were very stressed out. The manager of the store came to talk to Dad at the time, because you had said he was dishonest and that he ripped people off, and he wanted to assure Dad that that was not the case. We were very surprised you had said those things, because it was so unlike you. He didn't mention that those thoughts came from God or the Bible, though. All in all, we were a little concerned, but more because we thought you had thrown away a good opportunity to improve your life. I guess you could say we were disappointed.

I gave up my habit shortly after. I went through a trough of depression, which lasted several months. In December of the following year, I married my sweetheart. We had some good times and some bad times. We also had three children. Then, one night, she announced she had been having an affair with a man she had met in a nightclub, and that she was leaving me and our three young children. From the first, I was devastated. I thought I had lost the love of my life and would never know love again. Over a period of a couple of weeks, my mind broke. Mum, what were the first signs that I had crossed the line between sanity and insanity?

There is no one particular occasion or event that I would say really indicated you had lost your mind, but you were very distraught. There were tears, so many tears. From the moment Delta left, you began to change. From the outside, it seemed you slowly lost your mind over a period of weeks.

For a period of time, I stayed in the home we had been renting. I went for long walks alone, during which time I pleaded with God to bring my young bride back to me. My thinking became increasingly disorganised and chaotic. I began to experience thought insertions that seemed to arise from entities from outside of my mind. At various times I thought it was God, a demon or my wife. I took to lying on a mattress in the living area, listening to the entities talk to my brain. I came to know them as 'the voices', but I did not hear them as auditory hallucinations. They were confined to my thoughts. Distress drove me out of the rental house, and I bunked with friends. I became increasingly obsessed with religion, and talked about it constantly. Eventually, I ran out of friends to bunk with. I called my parents. Mum, do you remember the wet night you picked me up with the shopping bag that held my few possessions outside of a liquor store?

Vaguely. I remember thinking then already that you were delusional. You got in the car and started talking about God. Initially, I was glad that you'd turned to God for comfort, but I became aware very quickly that your mind was not as it should be.

I remember returning to my childhood home and collapsing on a mattress that was on the floor in what had been my old room. I didn't have a television, but I didn't need it. In fact, I was unable to watch television for several years after the night my mind broke. There were a number of reasons for this. First, the words seemed to be disjointed, and I would have to assemble every sentence consciously in my mind. Second, I was acutely aware of noises usually relegated to the background, like doors closing and fans blowing. Third, the dialogue and voiceover of the television shows became loaded with significance, and each line seemed like a message directly meant for me. The solution was very simple: no television. Instead, I conversed with the entities in my mind. At first, I did this directly with my thoughts, but as my mind became more chaotic, I began talking aloud. Mum, you once said that watching me talk aloud to the voices was one of the most difficult things you have had to do as a mother. Can you describe how you felt, and how you coped with it?

I'd heard voices several times myself, but it was always a case of did I really just hear that, or was it just my imagination? It was nothing like what you seemed to be going through. There was simply no Jason left anymore. You talked to the voices continually. It wasn't once a day, or twice a day, but all day, every day. It was very frightening. Perhaps the most frightening thing is you were steadily getting worse – you were getting louder, and more confused.

One thing I am personally curious about is whether the things I shared, and the manner in which I shared them, made much sense. Mum, was I coherent during this time, or was I impossible to understand?

Most of the time – I would say about ninety percent of the time – you were incoherent. I don't mean you were babbling nonsense, but you weren't making much sense, either. You might talk to me for half an hour – or rather, I would stand and listen for half an hour – and I'd still have very little idea of what you were actually trying to say. It wasn't like you went from one topic to the next, either. It was always about the same thing. It always came back to God talking to you. You said that you and He had a special friendship. Once, you said you were just like the prophet Samuel. You said you heard God calling, and you listened. And not just listened, but had conversations with Him. It was obvious from the things you were saying that you were drawing comfort from the Bible, but it was also obvious the Bible was something else to you. You would quote texts as if they were personal messages meant just for you. We were so, so confused. We didn't know what to think.

You know, there were also times when it seemed you were bent on destroying Delta. You used to say things – crazy things – about her, and I used to wonder if you wanted her dead.

At times you couldn't finish a conversation, or you would leap from one idea to the next without there being a sensible connection. When you couldn't finish a conversation, you would just walk away – sometimes to go pace outside – and then you'd be back. At first, I used to follow you, to encourage you to talk about it, but then you would forget what we were talking about. I wondered all the time whether a mind could possibly come back from that.

Dad found it very hard to listen to it when you were like that. I had some experience of what it was like to hear voices, so I encouraged you to talk about it to me, but Dad just couldn't cope with it.

Another thing Dad found very hard to deal with was the crying. You and I cried many times together. Sometimes, we would sit, holding hands, just crying. But at other times, it wasn't just crying – it was horrific, heart-wrenching wails. There was so much pain. Dad didn't know what to do, or how to fix it. There were even times, Jay, when I would turn the music up to full volume so we wouldn't have to listen to it. It was a horrible, horrible time. Dad and I often asked each other whether you would ever become well again. It wasn't until much later – when you were finally admitted into hospital – that we dared actually hope you might become well.

While I was in this frame of mind, I was driven to pace a lot. I also rocked compulsively. It was very uncomfortable for me to stay still, and I held a lot of muscular tension in my body, particularly in my jaw. Mum, did I adapt any other mannerisms that you thought were odd?

There was the terrible wailing, of course. It was an indescribable sound. Also, you'd often be sitting, and you would leap up, open the door, close it, and sit back down. We would ask, 'What's the matter, Jay?' and you would answer, 'Just waiting.' That was your response so often. At first we were confused, but then we kind of figured it out that in your head you were waiting for the moment when Delta would return. I'm not sure whether we were right about that, but that is the conclusion we came to.

The pacing was something else. You wouldn't just pace. You would march, like a man on a mission. Often you would be talking to yourself, and we would ask, 'What are you talking about, Jay?' and you would just stare back blankly, like the question didn't make any sense to you.

You were always stretching.
You'd stretch your arms, and your
legs, and move them in circles, trying
to shake the tension out of them. You
were always complaining they were
sore. You would also stretch your jaw
by opening your mouth really wide. It
was peculiar at first, but after a while,
we got used to it, and it became quite
normal to see you just stand on the
spot for a while, stretching.

My sister – Emily – was one of the
first to remark on my mannerisms, and tell
me I needed help. I seem to remember her
being there, then not being there, but I
don't remember her ever leaving. To this
day, I cannot be sure whether she
remained living with my parents when I
was unwell, or whether she went to live

elsewhere. My brother and my other sister lived elsewhere, and so missed the worst of my melancholy and madness. Mum, how did my brother and sisters take the news I wasn't well? Did they offer opinions at all on how to deal with me?

Emily was home the whole time you were there. She encouraged us to let you talk, even when you weren't making sense. She was very good to you while you were sick, and a good support for me. I am not sure how much you remember, but she used to go for long walks with you to give you someone to talk to besides the voices.

Your brother, Wayne, and your sister, Jenita, didn't visit much, but Wayne rarely visited during that time anyway. If you remember, he didn't have transport. I used to visit him, though, and we would talk about you. I think he stayed away a little more than he otherwise would have because he was a little unsure of what to say to you, or how to deal with you. He never stopped caring, though. Jenita lived in the north of the state at the time, and so didn't visit much either. We would talk on the phone about what you were struggling with. I think Jenita shut down a little when we spoke of it. Maybe that was one of the reasons they didn't travel down to see us very often.

One feature of my mind at this time is I began to hold some pretty heavy delusions. Delusions work a little differently than you might expect. In a rational mind, there is one reality. In a delusional mind, multiple realities may be held side-by-side. For instance, I believed there was a real Delta, but I also believed Delta was one of the entities in my brain. In fact, one of the purposes of talking to the entities was to convince Delta to come back to me. I did this by not only trying to persuade her, but by conducting my mind in such a way that she would respect it, just in case she was ever truly 'listening in'. One of the things I did to impress her was burning my forearms with cigarettes while keeping my mind free from fear and pain. Mum, do you remember me going through that stage? How did it make you feel?

At first I wasn't aware your burns were deliberate, because I saw only one or two, and you made believable excuses for them. Emily was the first to become concerned, and she made us aware that you were actually harming yourself with cigarettes. We were shocked, and talked about it with you. Every time you said it was the last time, and you wouldn't do it again, but every couple of days you would have a fresh burn mark on your arm.

I think one of the reasons you were hurting yourself was because your self-esteem was so low. After Delta left, you really seemed to think your life was over. When we talked about the burns, I got the impression that you somehow thought you deserved to be hurt.

At that stage, we tried to convince you to get help, but most of the time you thought that there was nothing wrong. The few times you actually admitted it was a problem, you told us there was nothing the doctors could do – it was because Delta left, and it wouldn't be made right unless she came back.

I remember the night my parents came into my room and sat to talk with me about what I was experiencing. I was sobbing, and I remember looking at my Dad, and in great jerky sobs saying, 'Dad, I hear voices.' I expected to see disbelief in his face, but instead, I saw compassion. It meant so much to me that he believed me.

The truth just seemed so incredible, so like something out of a horror novel, that I had expected him not to believe me. Do you have a recollection of that night, Mum?

Yes, I recall that night. We'd been watching you struggle for a while, and seen you go down, down, down. My recollection is not really so vivid, because I was just so distraught. I knew at that stage things were terribly wrong, and all I could think was that I'd somehow failed you, that this was my fault. I remember just crying and crying while Dad did all the talking. We talked about you getting help, but you insisted that God had chosen you for this, and that it would go away when He was ready.

One thing we have never done as a family is play the blame and shame game. No one has accused anyone of being the cause of my schizophrenia, and no one has caused me to be ashamed of it. Mum, as a parent, were there moments you wondered what you had done wrong to have me lose my mind?

Yes, all the time. The guilt started already when you were much younger. You were still a young boy when I was first admitted into hospital. Shortly after that, your youngest sister was born, and I thought I'd contributed something good to the family, and that it would all be alright. Then you started getting belly-aches before school, and I thought it was because you were insecure. At times, when you were younger, I had been really unreasonable, and I thought your insecurity came from that. I also thought that somehow your struggles with your mental health were caused by bad vibes I'd been sending out.

When you became really ill, I didn't know how to handle it. I thought that having been through a little of it, I was prepared to help you through anything, but your illness was something else entirely.

If you want the truth, it still bothers me to this day. I still ask myself what I could have done differently.

I imagine there were many tears cried behind closed doors. Mum, as a couple, how did you support each other while this was going on?

We used to pray with each other continually. Dad found it even harder to cope with than I did, because I had had my own experiences of mental health problems. I felt partially responsible, and very guilty, and Dad would continually assure me it wasn't my fault. Of course, Dad went to work during the day, so I would often be home alone with you. At the end of every day, Dad would be my sounding board. We talked about it all the time.

One day, I was walking down the main street, and I decided to stop in at a second-hand shop. They had a bunch of CDs on sale for a dollar, and I scrounged through them for artists I recognised. I discovered Eminem's *The Eminem Show*. I played it through, and then played it again. I found the themes a little confronting, but there was something about the rhyming words and rhythm of his voice I found soothing. I memorised the album, and when I wanted to silence the entities in my head, I would rap the lyrics. Mum, that album must have driven you out of your mind. Do you remember me playing loud music? I can't recall a single time you asked me to turn it down. Were you aware then that it was a coping strategy?

Oh, yes. Very aware. We did actually ask you to turn it down, though. Repeatedly. We often had to ask you to turn it down when something else was going on, or we had visitors. You used to say you had to play it loud to keep the voices away. It was always the same songs, the same tunes.

Sometimes you would talk while the songs played. It seemed like you were talking back to the voices that were rapping. I think sometimes you thought they were speaking to you.

Those songs, those tunes, got under our skin so much. I hope to never hear them again, to be honest.

During my crisis, I was – too some degree – oblivious as to how others were reacting to me. This was not always the case, though. I went through a period of several weeks when I would walk the ten kilometres into town every second night or so to engage strangers on the street in conversation. I have a distinct recollection of seeing pity and concern in their faces. At the time, I thought it was because my story touched them and that they cared, but in hindsight, they could probably see I was extremely unwell. However, in general, I was oblivious to how I was being perceived and how I was being treated. Mum, was there ever a time during my crisis when you felt the need to employ the 'tough love' approach? Did you ever try to 'snap me out of it'?

No, I don't recall ever doing that. Very subtly, I would encourage you to do things like play different music, but you didn't really listen. Other people told us not to put up with it, though. I don't know what they expected us to do. They would watch you as well, pacing and talking under your breath, and they would come up with strategies to help us to deal with it. We didn't do any of them, though. We thought getting tough would push you even more over the edge,

After eighteen months living under my parents' roof, I was granted government housing. I had been collecting household items such as furniture and white goods during this time, and was ready to move out. Mum, be honest. Did you think I would manage on my own?

That was a real source of worry. I am not sure how well you remember this, but for a while after you moved out, we visited every day. You weren't eating, and sometimes we would bring you meals. We were so scared you would hurt yourself, and that we would arrive to find you dead. We lived in fear all the time.

There were many times we would visit, and you would just burst into tears. Then you would just start talking. It never made a lot of sense. You used to ramble a lot about Delta, and about God, but it always seemed to come from that place in your head that only made sense to you.

Several months after moving into my unit, I started to develop pain in my hip. One night, the dominant entity in my head – I thought it was God – told me it was broken. The next day, I drove myself to the ER. An observant nurse noticed I was acting peculiarly. She called a mental health nurse. He assessed me, and immediately admitted me into hospital. Mum, were you surprised to hear they had admitted me into hospital for mental health reasons?

Surprised? No. Relieved? Absolutely! We thought help was finally on its way. We had confidence in the care you would get there because I had been in hospital, and I knew that was where you needed to be.

I was moved from a general ward to the psychiatric ward after a day or two. I remember a psychiatrist with a gentle, kindly manner asking me whether it was possible that the entities that spoke to me were not God, or demons, or my wife. She proposed that I had a form of mental illness called schizophrenia. I recall deciding to play along, and see where this took me. I didn't believe I was sick. It would be many months before I accepted that what I had was a condition that others have, and that I wasn't a prophet. During my time in the psychiatric ward – four long months – my parents visited me every day, and often brought my children to see me. Mum, that must have been exhausting. Wherever did you find the strength?

From above. We leaned on God heavily during that time, praying for strength and comfort. That, and our love for you and your children saw us through.

After I was discharged from the hospital, I returned to my parents' house. Again, there was no blame or shame. I was a little like a zombie due to the powerful doses of antipsychotic drugs I had been prescribed, and twelve sessions of electroconvulsive therapy. Mum, during this time, did you ever fear I would never regain the full use of my mind?

Yes. For a time, there was no indication that your mind would ever get better. It took a long time – months and months – before we began to see even a glimmer of hope. Those were awful times.

Even after that, there was a barrier when it came to talking to you. You spoke about sensible things, things you had read or seen on television, but you just were not good at communicating them. You would start talking, and you'd lose me a couple of words in. I simply nodded and let you talk.

In the last ten years, I have come a long way. Part of the healing is down to finding words for my experience. Writing and talking about my condition has been my therapy. I often talk about it with my mother, and she listens patiently, and occasionally gives me advice. Mum, do you have any final advice for people caring for someone with schizophrenia?

Pray. Pray and work. Seek medical advice as soon as possible. Don't just think that tomorrow will be a better day – work at making tomorrow a better day by seeking help. Talking is very important. Don't stop sharing, or it will snowball.

www.ingramcontent.com/pod-product-compliance
Lightning Source LLC
Chambersburg PA
CBHW072301170526
45158CB00003BA/1135